The Learning Works

Designed and edited by
Sherri M. Butterfield

Copyright © 1993
The Learning Works, Inc.
Santa Barbara, California 93160

Library of Congress Catalog Number:
92-074100
ISBN 0-88160-219-1
LW 303

Printed in the United States of America.

Current Printing (last digit):
10 9 8 7 6 5 4 3 2 1

Introduction

SUPERDOODLES are books that provide simple, step-by-step instructions for super line drawings. The animals in this book may be sketched large for murals or posters, or small for bookmarks and flip books. They may be used individually in separate pictures or combined to create a desert, forest, or jungle scene.

As you follow the steps, draw in pencil. Dotted lines appear in some steps. Make these lines light so that they can be easily erased later. When you have finished your drawing, erase all unnecessary lines. To give your drawing a finished look, go over the remaining lines with a colored pencil, crayon, or felt-tipped pen.

If you enjoy this book, look for other **Learning Works SUPERDOODLES**. Titles in this series include *Dinosaurs, Rain Forest, Sports,* and *Vehicles.*

African elephant

An elephant may eat 500 pounds of plants in one day. Draw an elephant that is using its trunk to pull leaves and branches from a tree.

armadillo

Armadillos eat ants, termites, and other insects. Draw an armadillo licking up ants with its long, thin, sticky tongue.

bison

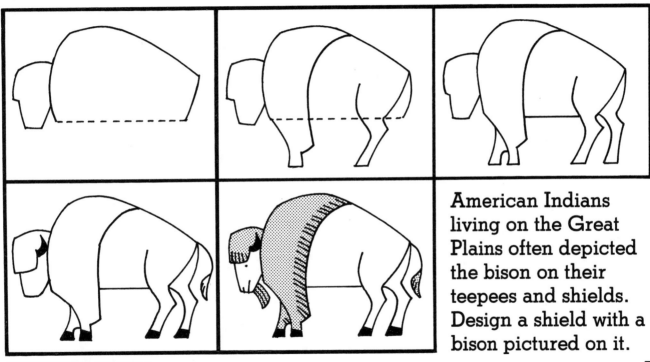

American Indians living on the Great Plains often depicted the bison on their teepees and shields. Design a shield with a bison pictured on it.

cheetah

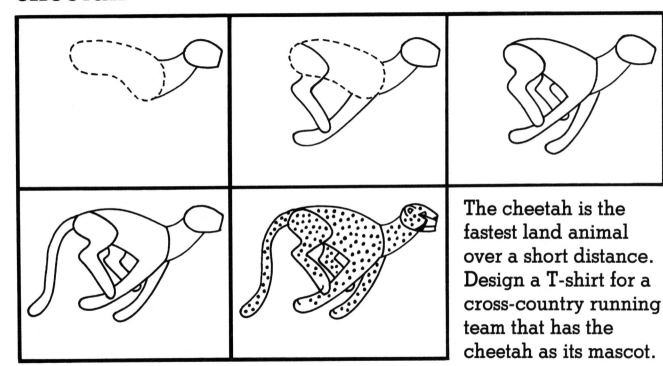

The cheetah is the fastest land animal over a short distance. Design a T-shirt for a cross-country running team that has the cheetah as its mascot.

chipmunk

Chipmunks carry home acorns and other seeds in their cheeks. Draw a chipmunk in its underground burrow surrounded by acorns and nuts it has gathered.

7

coatimundi

These animals use their long, rubbery snouts to dig up the worms and bugs that they eat. Add some food for your coatimundi to find.

douroucouli

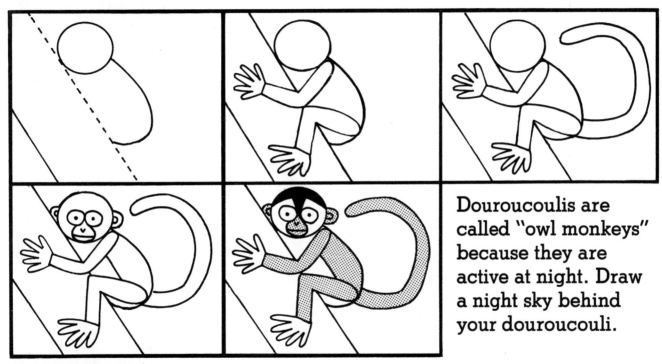

Douroucoulis are called "owl monkeys" because they are active at night. Draw a night sky behind your douroucouli.

dromedary

Dromedaries are still used for transportation in some African and Asian desert areas. Give your camel something to carry.

elephant shrew

These wiggly nosed creatures may be only 4 or 5 inches long. Behind your shrew, draw a leaf that is larger than this tiny mammal.

flying squirrel

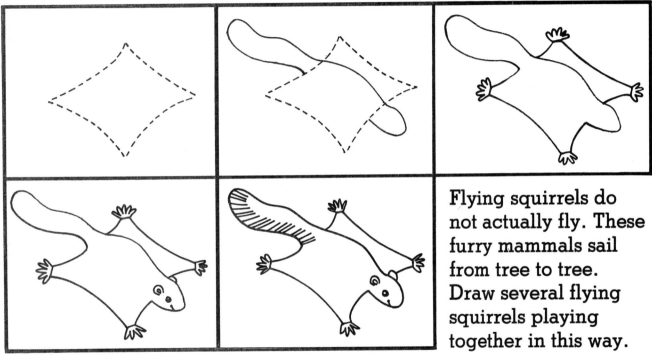

Flying squirrels do not actually fly. These furry mammals sail from tree to tree. Draw several flying squirrels playing together in this way.

galago

Galagos live in rain forests. They jump easily from tree to tree, catching insects to eat. Put a moth in your galago's paw.

giant panda

Pandas are playful! Roll your panda over on its back by turning your drawing upside down. Put a big ball between your panda's front paws.

gibbon

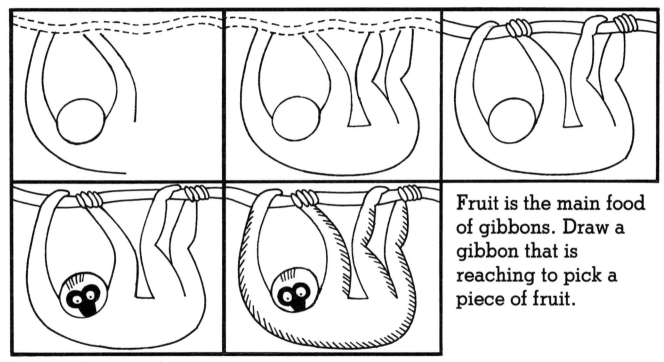

Fruit is the main food
of gibbons. Draw a
gibbon that is
reaching to pick a
piece of fruit.

gorilla

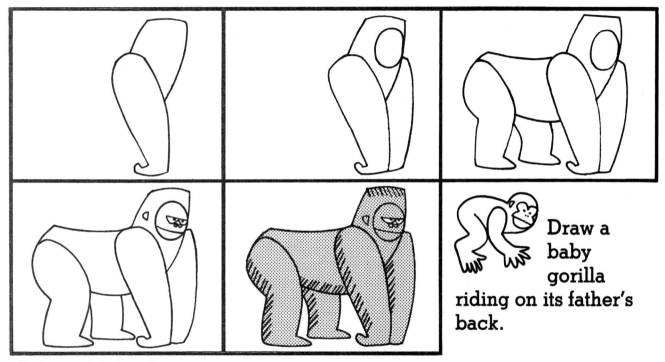

Draw a baby gorilla riding on its father's back.

horse

Put a rider on your horse and draw a field, fence, or forest in the background. You may wish to give your horse spots or other markings.

ibex

Male ibexes compete for herd leadership by butting their heads together, trying to knock each other down. Draw two male ibexes fighting.

SUPERDOODLES: MAMMALS
©1993—The Learning Works, Inc.

18

Indian rhinoceros

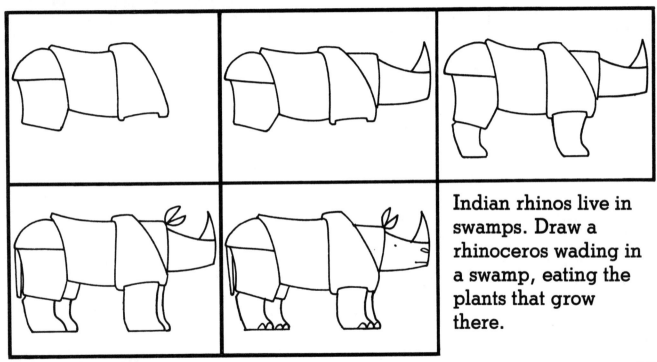

Indian rhinos live in swamps. Draw a rhinoceros wading in a swamp, eating the plants that grow there.

jackrabbit

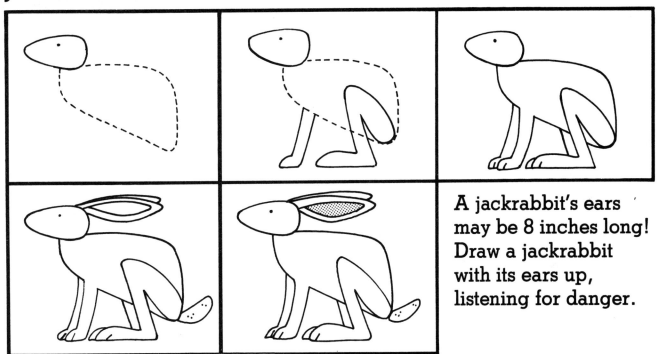

A jackrabbit's ears may be 8 inches long! Draw a jackrabbit with its ears up, listening for danger.

kangaroo

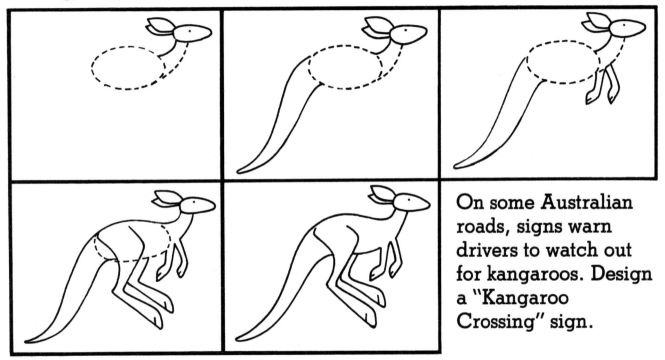

On some Australian roads, signs warn drivers to watch out for kangaroos. Design a "Kangaroo Crossing" sign.

kangaroo rat

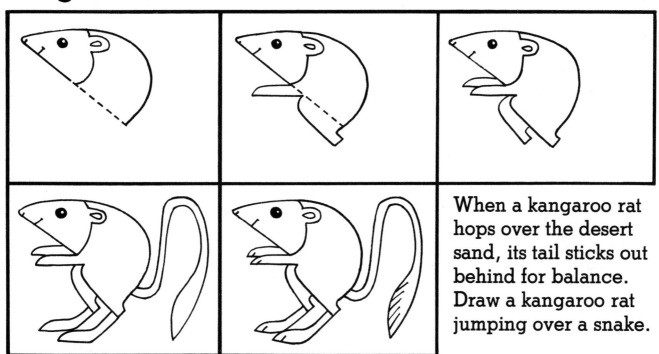

When a kangaroo rat hops over the desert sand, its tail sticks out behind for balance. Draw a kangaroo rat jumping over a snake.

koala

Koalas live in small groups. They rarely leave the treetops. Draw three koalas hanging onto the same tree trunk.

long-eared bat

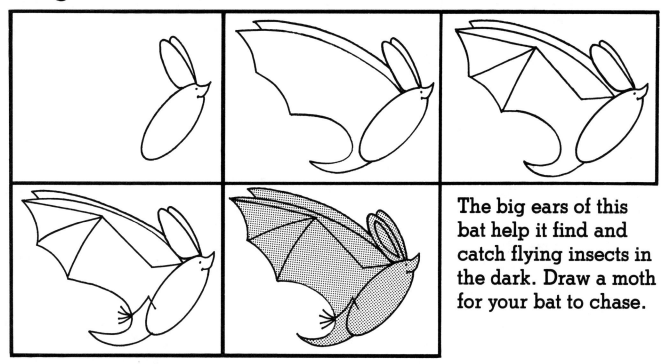

The big ears of this bat help it find and catch flying insects in the dark. Draw a moth for your bat to chase.

moose

The moose likes to feed on water plants. Draw a moose wading in a pond, eating lily pads and cattails.

orca

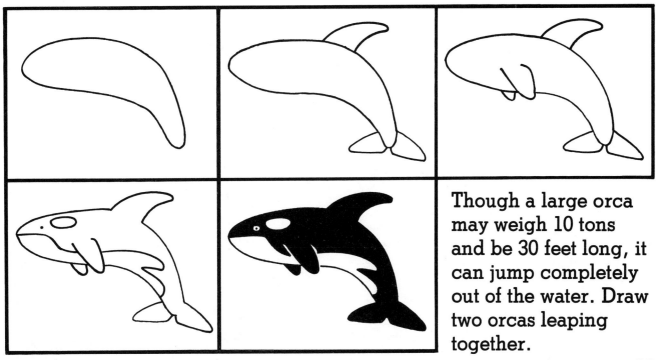

Though a large orca may weigh 10 tons and be 30 feet long, it can jump completely out of the water. Draw two orcas leaping together.

raccoon

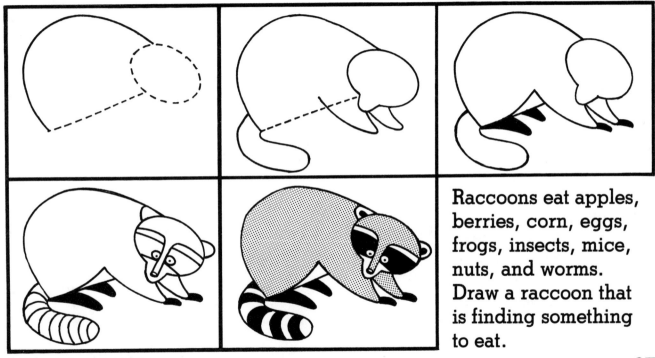

Raccoons eat apples, berries, corn, eggs, frogs, insects, mice, nuts, and worms. Draw a raccoon that is finding something to eat.

red fox

Foxes like to run and play in the snow. Draw a fox trotting through the snow, leaving a long line of tracks.

ruffed lemur

The African forests where lemurs live are being destroyed. Make a poster that asks people to help protect this habitat.

springbok

The springbok is known for its stiff-legged leaps, which lift it as high as 10 feet in the air! Draw a springbok leaping over a person's head.

warthog

Warthogs like to sleep in holes dug by other animals. Draw a warthog sleeping in a large, shallow hole.

walrus

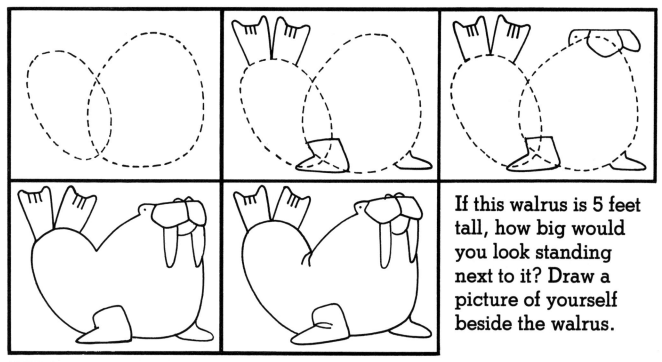

If this walrus is 5 feet tall, how big would you look standing next to it? Draw a picture of yourself beside the walrus.